Solar Photovoltaic Resource for Residential, Commercial and Utility Systems

Edition 1

Cover Picture: Broken cloud days can cause surges and drops in power from solar photovoltaic systems

Contents

1. Introduction

Solar photovoltaic technology has now become mainstream and is widely used around the world. All of the items needed to construct a reliable system can now be purchased from many different vendors and standards now exist in the industry to ensure that these different products can be seamlessly integrated together. The future of solar photovoltaics is bright and rapid adoption of the technology is underway.

This book is a collection of notes, diagrams, pictures and charts all in one place for those who are involved in the solar photovoltaic field. You will find it to be the ideal resource of information when you need to find out something quickly.

Solar photovoltaics is very different from conventional electrical system theory and because of this it has its own electrical codes that are contained in National Electric Code (NEC) Section 690 Solar Photovoltaic Systems in the USA. This book is to be used in conjunction with the National Electric Code (NEC) for residential and commercial installations and also the National Electric Safety Code (NESC) for utility installations. For construction of systems it should also be used with the local building code books.

2. The Basics of Solar Photovoltaics

Solar photovoltaics comes in many different types:

- Monocrystalline
- Polycrystalline
- Thin film
- Other technologies

Although the technology is different between each type, they all do the same thing. All direct current (DC) solar modules generate DC electricity when exposed to sunlight. The listing above is in order of how efficiently each type will convert sunlight into electricity.

A typical direct current (DC) module will have the following electrical ratings on its label:

- Temperature adjustments
- DC Open circuit voltage (Voc)
- DC Maximum power point voltage (Vmpp)
- DC Short circuit current (Isc)
- DC Maximum power point current (Impp)
- DC Rated system voltage

All of these values are given for Standard Test Conditions (STC). Lets look at what each one of these mean:

Standard Test Conditions (STC)

Standard Test Conditions (STC) is how the solar module performs at a temperature of 25 °C, an irradiance of 1000 W/m^2 with an air mass 1.5 (AM1.5) spectrum. This is a standard test for all solar modules that are manufactured for the USA market that was developed by the photovoltaic industry and the government. It represents an average set of conditions that can be expected at the mid point between North and South of the contiguous forty-eight states during spring and fall with the sun perpendicular to the solar module. San Francisco, California and Wichita, Kansas are near this midpoint of 37 degrees latitude. In Asia Seoul, Korea, is near and in Europe both Sevilla, Spain and Cantania, Italy are near to 37 degrees latitude.

It is important to note that a solar module output will be continuously variable during the year and even during the day. In wintertime it will output less power than its rating and in summertime it will frequently output more power. These electrical ratings are for guidance only and it is where many new photovoltaic designers make mistakes in thinking that the module will never output more power than its rating. It is important that you understand that these solar photovoltaic modules can output far more power than their label states. It can be over fifty percent more and this will need to factored into the system design.

Temperature Adjustments

Solar modules are affected by temperature, both hot and cold, and adjustments to the module ratings needs to be made for the operating temperature outside of 25 degrees Celsius. It is important when designing a system that the historical temperature minimum and maximum values are known for the area where the system is to be installed and these adjustments are factored into the design.

Open Circuit Voltage (Voc)

The open circuit voltage rating is how much voltage the module will put out with no load attached. This is an important value in order to design a system. This is the voltage to use when selecting your components and it must be adjusted for the historical minimum and maximum temperatures for the area. If more than one module is connected in series then multiply this temperature adjusted voltage by the number of modules in series to get the total maximum DC voltage of the system.

DC Maximum Power Point

The DC maximum power point is a simple concept. Power is a function of both voltage and current. The maximum power point is obtained when the current and voltage from the module when multiplied together give the maximum power figure. These values will change constantly during the day with the weather conditions. Voltage will remain relatively constant, but current will vary a lot with irradiance. The DC to AC inverter system constantly monitors the power from the solar photovoltaic DC system and automatically keeps the inverter system working at the maximum power value for the given conditions.

DC Maximum Power Point Voltage (Vmpp)

The DC maximum power point voltage (Vmpp) is the operating voltage of the solar module under load. Again this value will change with temperature and irradiance, but should only vary by about twenty percent of the STC rating during the day time.

DC Short Circuit Current (Isc)

The DC short circuit current value is the maximum current that the module will output at standard test conditions if the positive and negative terminals were connected (shorted) together. It is important to note that this value will vary a lot dependent on weather conditions and can be over fifty percent larger during summertime.

DC Maximum Power Point Current (Impp)

The DC maximum power point current is the amperage that the solar module will output at standard test conditions in normal operation. It is important to note that this value will vary a lot dependent on weather conditions and can be over fifty percent larger during summertime.

DC Rated System Voltage

This is a very important design value. It is a rating of how many modules can be safely connected together in series, this is called a solar photovoltaic module string. This system voltage should never be exceeded when adjusting for the minimum and maximum temperatures of the area that the system is being installed. This value basically limits the number of modules that can be connected in series in the system.

3. Photovoltaics and Weather

The performance of any solar photovoltaic system is dependent on the weather. The main factors that affect the system performance are clouds, irradiance, temperature, shade, latitude and how dirty the solar modules are. Let's now explore the effects of the weather in more detail:

Irradiance

Irradiance is a measure of how much sunlight the solar module is receiving. It is given in watts per meter squared or W/m^2. Standard Test Conditions (STC) uses a value of 1,000W/m^2. This value can range from 0W/m^2 at night through to over 1,500W/m^2 during a day interspersed with large fluffy clouds. This value of 1,500W/m^2 is larger than what you would receive in space. The reason why we can get greater values at ground level is due to what is known as the "cloud effect". Normally the sunlight is traveling in a straight line from the sun to our solar module with some atmospheric scattering. However, when clouds are present they can also reflect and can act like lenses to send some extra sunlight onto the solar modules. This extra light is converted into extra energy and this is seen largely as an increase in power from the system. This effect can be a few minutes long in duration when it occurs.

The diagram on the next page demonstrates the "cloud effect".

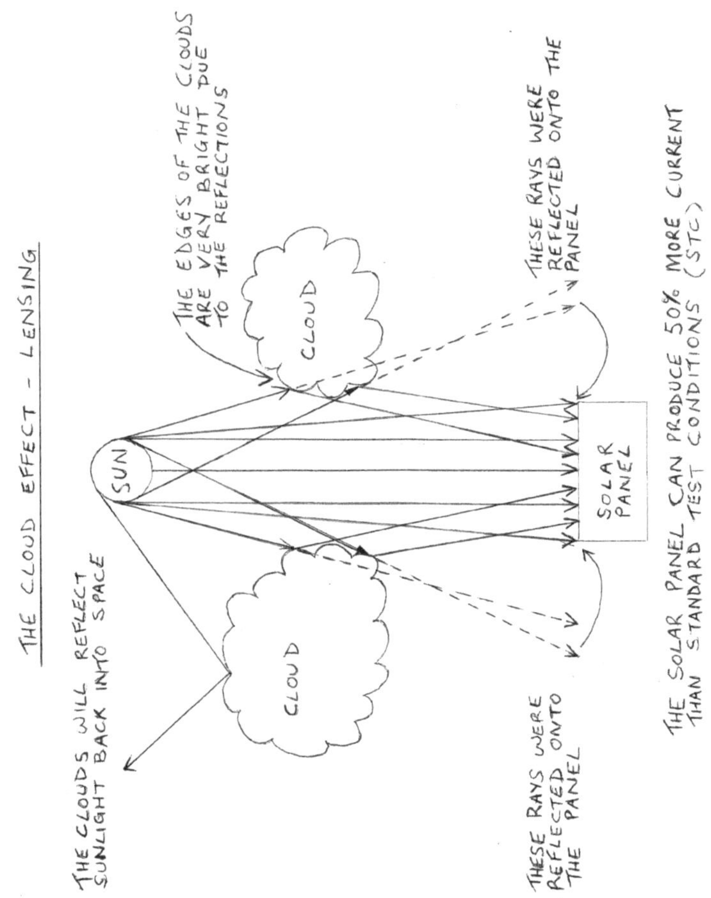
THE CLOUD EFFECT - LENSING
THE CLOUDS WILL REFLECT SUNLIGHT BACK INTO SPACE
THE EDGES OF THE CLOUDS ARE VERY BRIGHT DUE TO THE REFLECTIONS
SUN
CLOUD
CLOUD
SOLAR PANEL
THESE RAYS WERE REFLECTED ONTO THE PANEL
THESE RAYS WERE REFLECTED ONTO THE PANEL
THE SOLAR PANEL CAN PRODUCE 50% MORE CURRENT THAN STANDARD TEST CONDITIONS (STC)

Other effects on irradiance are the snow effect, the water effect (lake/ocean/wet surfaces after rain), the building effect and albedo. Snow cover, water, glass covered buildings, reflective painted buildings and roofs, and the albedo of the area surrounding the solar photovoltaic modules can reflect extra sunlight onto the solar power system. If you are installing a system in an area that has any of these, it is important to account for it. Each effect can produce an increase in power output. If you find yourself having to wear sunglasses in the solar photovoltaic system location for your eyes to be comfortable, then you probably have light reflections taking place.

In wintertime generally the system will operate at below the standard test conditions values and in summertime it will generally exceed these values. During the design phase of the system you will need to assess where the greatest power need is and perhaps increase the size of the system accordingly if it is in wintertime.

Air Mass

Air mass is a measurement of the the amount of atmosphere that the sunlight has to pass through to get to the ground. It varies with the seasons and also the location on the earth. Within the tropics, air mass will reach its maximum power value of 1 during summertime. Air mass 1 corresponds to the sun being directly overhead, air mass increases as the sun moves from directly overhead down to the horizon.

All USA solar modules are rated for air mass 1.5 which corresponds to a central USA location in Spring and Fall. When in a southerly location you will approach air mass 1 which will increase power output by about 13% from STC in the USA.

Locations that are at or near air mass 1 in Summer time in the USA are all Hawaiian islands, Florida and Texas. Approximately half of the continental USA is located between air mass 1 and air mass 1.5 in Spring and Fall. If you are working on systems that are located in these Southern USA states, you will get more power out of these systems due to a decreased air mass. In summertime the air mass will move closer to 1 in the continental USA.

Clouds

Clouds come in many forms. An important question is how do clouds affect irradiance on solar power photovoltaic power systems? The list below will help with understanding the effects of clouds on irradiance at air mass 1 (within the tropics in summer time):

- Clear, sunny skies will give approximately 1,130W/m^2. The transmission characteristics of the atmosphere will vary in clear skies, sometimes being relatively transparent and other times being more opaque and this affects irradiance values. Air quality is a major factor for the transmission of sunlight through the atmosphere. Particulate matter in the atmosphere will reduce the transmission level.
- Thin cirrus will give approximately 1,000W/m^2. Thin cirrus will give even and relatively stable irradiance levels due to scattering of the light.
- Thick cirrus will give approximately 750W/m^2.

- Thin clouds will give about 500W/m^2.
- Thick clouds will give about 250W/m^2. No shadows on the ground will be present
- Thick clouds with a visibly dark sky will give about 100W/m^2. No shadows on the ground will be present. You will not be able to see the location of the sun in the sky.
- Tall and dense broken clouds will give surges of about 1,500W/m^2 and reductions to about 100W/m^2 of irradiance due to the cloud effect. The rate and length of time for these surges and reductions is dependent on the speed of the clouds passing in front of the sun.

Temperature

Temperature will affect the system to a much lesser extent than irradiance. The cooler the system is below 25 degrees Celsius, the more power it will produce. Correspondingly, the hotter the system is above 25 degrees Celsius, the less power it will produce. Temperature can affect solar photovoltaic systems power output by about twenty percent.

Shade

It is undesirable to shade solar photovoltaic modules as it can significantly affect the performance of the system. When studying the location of where to install a system, always factor in the surroundings for shading effects. Avoid shading with solar photovoltaic power systems.

Wind

Wind will provide cooling to the photovoltaic modules and it is an aid to power production. A breezy location will provide improved performance from the system. When mounting solar modules onto racking, it is good to allow spaces between the solar modules in order to aid with cooling airflow around the modules and also to reduce wind resistance. When choosing solar modules and mounting systems, it is important to ensure that they are rated for the wind speed of the area that you are installing them into.

Altitude

A higher altitude location will improve the amount of irradiance that the system will receive, due to less scattering and absorption of the sunlight by the atmosphere. It also acts as a natural cooler of the system which further improves system performance. Generally a high altitude location will have a higher percentage of clearer skies during a year which will give a higher energy yield from the system.

Snow and Ice

Snow and ice should not affect a solar module, other than obscuring its view of the sun. Tracking systems can be affected by this and in some snowy locations it is advisable to park the solar system facing South during these periods. The reflection from the snow will increase the power from the system in Winter time.

Hail

Hail can break solar modules, so it is important to know type of hail that your area can receive. If you get large golf ball size hail, you may not want to install glass solar modules. Solar modules are tested for hail and pass the tests even if the glass module breaks. The test just ensures that the modules remains intact when broken. Glass solar modules are hard to break and normal sized hail should have no effect.

Dirt

Clean solar modules are the desirable configuration for a system. However, dust and dirt will get onto the surface of the modules and will degrade performance by up to 10% on average. Cleaning the modules is very much a function of the location where they are installed and also how dirty they are. Most people will clean on an as needed basis, generally when they are visually very dirty. Always follow the manufacturers instructions for cleaning your particular modules and remember that solar modules are operating with electricity flowing in them when exposed to light. Night time cleaning is recommended for safety.

Lightning

Lightning can affect solar modules, especially on large systems that cover fields. Good equipment grounding is the way to deal with this threat. A low resistance ground will generally dissipate lightning away from a solar module that is struck by lightning. Generally, the damage should be limited to only the solar module that was struck. If a cable is struck, then lightning surge arrestors can limit the damage in the system. These are generally installed in the inverter and on larger systems, in combiner and re-combiner boxes. Lightning may blow the string fuse/circuit breaker(s) for the module(s) struck. Install lightning protection as recommended by the manufacturers of the products used in the installation.

Seasons

We have four distinct seasons of Winter, Spring, Summer and Fall. We can word this another way as Winter Solstice (December 21), Spring Equinox (March 20), Summer Solstice (June 21) and Autumn Equinox (September 22). What does this mean to a solar power system?

- The length of the day
- The angle of the sun (air mass)
- Heating and cooling
- Rain

Winter solstice is the shortest day of the year and summer solstice is the longest day of the year. Spring and fall equinoxes are when day time is the same length of time as night time.

Regarding the angle of the sun in the USA, Winter Solstice is when the sun is at the lowest in the sky, or 23.5 degrees below the equator and Summer Solstice is when it is 23.5 degrees above the equator. Spring and Fall equinoxes are when the sun is directly overhead at solar noon at the equator.

For our solar power system, this means that we will produce our largest voltage in wintertime when it is the coldest and we will produce our largest current when it is summertime with peak irradiance.

The changing seasons will affect rainfall and in dry seasons you may want to schedule cleaning to keep the modules in good performance. Rain generally helps to keep the module clean naturally.

There are a number of things to consider with the seasons:

- Spring & Autumn
  - The system will be operating close to standard test conditions (STC) and measured values should be close to that on the solar module label.
  - This is the most favorable time for outdoor working.
- Summer time
  - The system will be hot and the DC voltage will be lower than normal.
  - The system DC electrical current will be at the highest value for the year.
  - Ambient temperatures will be high.
  - Heat and dehydration may be a problem for working on the system.

- Wintertime
  - The system DC voltage will be at the highest value for the year
  - The system DC current will be lower.
  - It may be too cold to work on the system
  - Frost, ice and snow may be an issue for performing maintenance.
  - Ambient temperatures will be low.

Due Diligence

It is important when designing, operating and maintaining a solar power generation system that you are aware of the annual climatic conditions to expect. Amongst the data that you should have is:

- Historic annual minimum temperature
- Historic annual maximum temperature
- Historic annual maximum wind speed
- Historic annual snow fall depth
- Historic annual hail size
- Historic annual peak irradiance
- Historic monthly irradiance

With these values you will be able to make educated engineering decisions regarding the selection of your system.

4. USA Wind Speed Zones

Wind speed ratings are important in solar photovoltaics due to the large surface area of the solar photovoltaic panels.

Most tracking systems have wind speed senors that will automatically park the modules in a horizontal position during high winds. If this sensor fails, then a tracking system may be destroyed in high winds, so it is important to keep a check on the wind speed sensor if it has this.

Other tracking systems are manually operated and need to be parked by the operator for high wind speed events. Keeping a check on the weather forecasts is needed for these systems.

Fixed tilt systems and inclined single axis trackers present the largest solar photovoltaic module surface area and will need a strong support system rated for the high wind speed rating of the area they are installed in.

The most important wind speed rating is that of the solar photovoltaic glass module surface. It must be rated for the wind speed as mounted.

The highest wind speed in all fifty states is 150 MPH and any solar photovoltaic modules and systems with this rating can be installed in any state.

The major problem for glass solar photovoltaic modules is surviving the flying debris that may hit the glass in a high wind speed event. Expect increased solar module damage after a high wind speed event if flying debris occurred.

The wind speed that systems should be rated for in the USA can be obtained from the relevant building code regulations for the installed system location. It varies widely by location.

5. Irradiance Pictures

The following pages show pictures of the different irradiance levels to expect in the field in summertime at air mass 1 and these were all taken with the same camera. Irradiance can cause problems with systems if these effects were not accounted for and designed into the system. Just as a reminder, irradiance and solar module current are generally proportional to each other. Increased irradiance will produce a corresponding increase in solar module current. There is no upper limit on solar module current output other than the string fuse blowing which is usually sized to be at least 156% of Isc.

Irradiance	Current (Isc STC)
2,000	200%
1,750	175%
1,500	150%
1,250	125%
1,000	100%
750	75%
500	50%
250	25%
100	10%
0	0%

Clear, sunny skies will give approximately 1,130W/m^2.

Thin cirrus will give approximately 1,000W/m^2.

Thick cirrus will give approximately $750W/m^2$.

Thin clouds will give about $500W/m^2$.

Thick clouds will give about 250W/m^2.

Thick clouds with a visibly dark sky will give about $100W/m^2$.

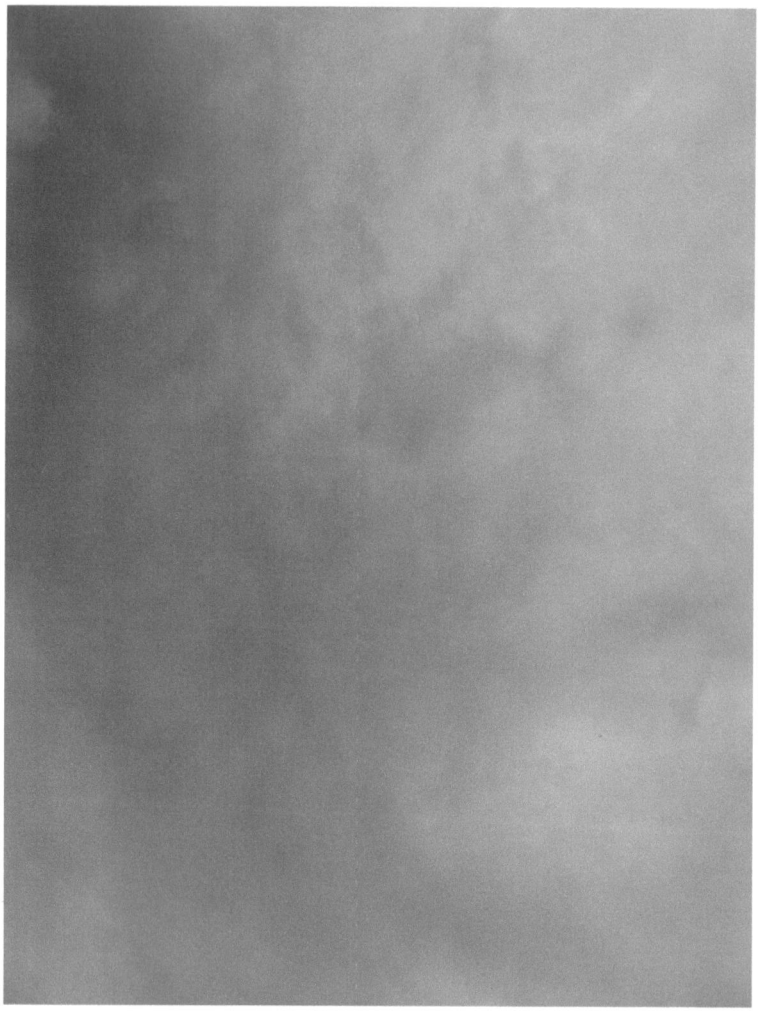

Tall and dense broken clouds will give surges of about 1,500W/m^2 and reductions to about 100W/m^2 of irradiance due to the "cloud effect".

6. System Selection

System selection is the most complicated stage of solar photovoltaic system designs and will determine how well your system works. When done well, your system will give years of trouble free operation.

Solar Modules

Solar modules typically represent the largest investment in the system. Traditionally silicon solar modules have been used, but now many new types of technologies are emerging such as thin film and so on. All solar modules are tested to the Underwriters Laboratory (UL) 1703 standard in the USA.

Silicon is the best understood, the most efficient and has been around for many decades. The newer film type solar modules are less efficient and cheaper to purchase. Unfortunately more thin film modules are needed to generate the same power as silicon, and this increases the system physical size and associated support systems such as cabling, racking, installation costs and so on.

Generally any decision on which technology to use is driven by market rates for each type of technology, aesthetics and personal preference. Solar modules are a commodity and their prices can fluctuate rapidly.

Mounting System

There are three ways to mount your modules:

All have their pros and cons.

The fixed tilt system is the most common and is widespread. The solar modules are either mounted to a roof, building or are ground mounted in a fixed position inclined to face south at a tilt angle matching the latitude. Some systems allow you to adjust the tilt angle of the modules for the season, but it appears that most people prefer the low maintenance option of mounting the modules into a fixed position for the entire year. The fixed tilt system is the most reliable configuration and also the lowest cost. The downside is it has the lowest annual energy output of the mounting systems.

The single axis tracker works well. The solar modules are mounted on a rotating North-South axis which allows them to track from East to West during the day. There are two types of single axis trackers generally available. The first has the North-South axis mounted horizontal and the modules can track in the East to West direction. This system works well in or near to the tropics where the sun can be almost directly overhead. The second has the North-South axis inclined to match the latitude and this enables the solar modules to face the sun in spring and fall. This system works better as you move further away from the tropics. A single axis tracker can increase annual energy output by about 24% when compared to a fixed tilt system. The single axis tracker does not cost much more than a fixed tilt system and the extra expense is generally offset by the extra annual energy yield of the system. The downside is that the tracking mechanism does need regular maintenance and occasionally will break down.

The dual axis tracker has the modules tracking the sun from sunrise to sunset, keeping the solar modules in the optimal position for maximum power generation. A dual axis tracker can increase annual energy output by about 30% when compared to a fixed tilt system. The downside to a dual axis tracker is that it requires a lot of space, can be very tall, has a complicated control system, they are expensive and they are the highest maintenance system.

Approximate adjustments to annual energy output using a South facing fixed tilt system inclined at latitude as the base value.

- 1.24 = Latitude inclined single axis tracker
- 1.3 = Dual axis tracker

The diagram on the next page shows the differences for each tracker system at noon with the seasons. Note how the dual axis tracker system always presents the solar modules with their maximum surface area to the sun.

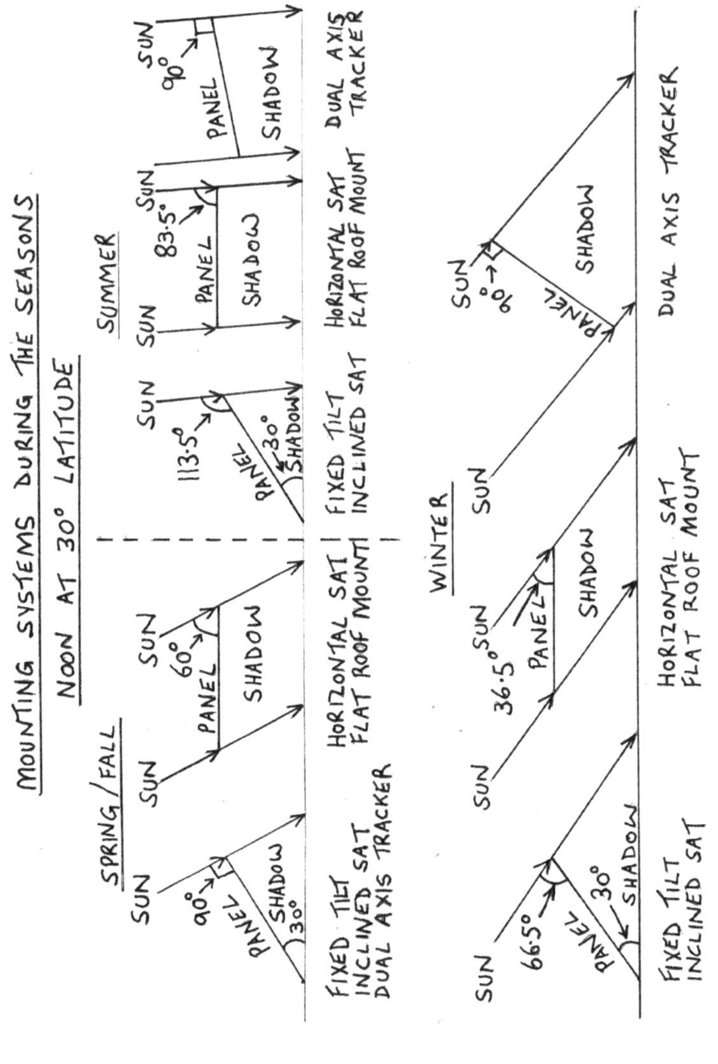
MOUNTING SYSTEMS DURING THE SEASONS
NOON AT 30° LATITUDE
SPRING/FALL
SUN
90°
PANEL
SHADOW
30°
FIXED TILT
INCLINED SAT
DUAL AXIS TRACKER
SUN
SUN
60°
PANEL
SHADOW
HORIZONTAL SAT
FLAT ROOF MOUNT
SUMMER
SUN
113.5°
PANEL
30°
SHADOW
FIXED TILT
INCLINED SAT
SUN
SUN
83.5°
PANEL
SHADOW
HORIZONTAL SAT
FLAT ROOF MOUNT
SUN
90°
PANEL
SHADOW
DUAL AXIS
TRACKER
WINTER
SUN
66.5°
PANEL
30°
SHADOW
FIXED TILT
INCLINED SAT
SUN
SUN
36.5°
PANEL
SHADOW
HORIZONTAL SAT
FLAT ROOF MOUNT
SUN
90°
PANEL
SHADOW
DUAL AXIS TRACKER

Cabling

For all outdoor cabling exposed to the sun, solar rated cable should be used due to its resistance to ultraviolet radiation. Once inside a building, conduit or underground this requirement need not apply.

On ground mounted installations it is important to remember that animals will be able to access the cables and equipment, so you will need be familiar with your local wildlife in the area. Where possible, enclose all accessible cabling with protective covers.

Conduit or ducting is recommended for all underground runs as it can be easily rewired to a larger size if you find that you system is generating more power than designed for or if a cable goes faulty.

Inverters

Inverters come in many different types and sizes. This book is dedicated to grid tie inverters and this is what we will consider. All residential and commercial solar photovoltaic grid tie inverters have to comply with UL1741 in the USA. This ensures that all grid tie inverters meet the requirements of the utility authority. Some of these requirements are below:

- Disconnect from grid during power cuts
- Protection against faults
- Good power quality output

Solar Design Hints and Tips

All DC Grid Connected Solar Photovoltaic Systems Need

- Equipment Ground.
- DC Disconnect.
- Inverter.
- AC Disconnect.

As They Get Larger They Also Need

- Combiner box.
- Recombiner boxes.
- Transformers and switchgear.
- Distribution and transmission systems.

Important Guidelines for Tracking Systems

- Try to put no more than one tracking system on one inverter.
- Install blocking diodes in each string to prevent reverse current flowing on multiple tracking systems installed on one inverter.

DC component selection

- Always make sure that DC rated components are used.
- Components must exceed specifications for maximum voltage.
- Components must exceed specifications for maximum continuous current.
- All outdoor cables must be solar rated.
- All outdoor connectors must be solar rated.
- De-rate equipment, cables and fuses for installed conditions such as:
    - High ambient enclosure temperatures.
    - Burial.
    - Enclosed conduits and ducts.
    - Heating effects.
    - Cooling effects.
    - High power cycling.
- Cables, breakers and fuses:
    - At least 156% larger than the short circuit solar photovoltaic current.
    - Increase current as needed for reflections.
    - Never exceed the interrupt ratings of the fuses and breakers.
    - Apply de-rating for highest expected fuse ambient temperature per fuse manufacturers de-rating tables.

  - Use solar photovoltaic rated fuses in the DC circuit.
  - Never exceed the string fuse size as recommended by the solar photovoltaic module manufacturer.
- Use metal enclosures for effective heat dissipation from the electrical equipment.

AC Component Selection

- Always make sure that AC rated components are used.
- Components must exceed system specifications for maximum voltage.
- Components must exceed system specifications for maximum current.
- All outdoor cables must be solar rated.
- All outdoor connectors must be solar rated.
- De-rate equipment for installed conditions such as:
  - High ambient enclosure temperatures.
  - Burial.
  - Enclosed conduits and ducts.
  - Heating effects.
  - Cooling effects.
  - High power cycling.
- Cables, breakers and fuses:
  - At least 125% larger than the maximum AC current.

- Apply de-rating for highest expected fuse ambient temperature per fuse manufacturers de-rating tables.
- Never exceed the interrupt ratings of the fuses and breakers.
- Use metal enclosures for effective heat dissipation.

Enclosure Mounting

- Exterior equipment locations should always be in the shade.
- All electrical equipment is to be above the flood plain

Inverters

- More inverters improves maximum power point tracking on the system
- Keep inverter power down to about 250 kW AC maximum per single inverter.
- Try to keep below 100 strings per inverter.
- If possible, mount inverters in shaded locations, consider constructing a shade canopy if needed.
- If inside a building or structure, ensure that the indoor ambient temperature can never exceed the inverter ambient temperature ratings.
- Use proven inverter technology for your system size.

Manufacturer Instructions

- Always design your system in accordance with the manufacturers installation manuals
- Follow the maintenance schedules in order to maintain warranty coverage

Codes

- Always design to local photovoltaic electrical codes.
- Always design to local building codes
- Consult with qualified engineers
- Have qualified engineers approve designs
- When in doubt, engineer on the side of safety.

Labeling

- Make sure all equipment is labeled in accordance with the local codes and the manufacturers instructions.

Safety

- Make sure all system and equipment grounding is installed correctly
- Make sure all covers are installed

- Make sure DC polarity of cabling is correct
- Make sure AC phase rotation is correct for three phase systems.
- Follow the local safety codes
- Never wash or touch grid interconnected solar photovoltaic modules by hand during the day time, as they may have up to 1,000 volts DC on them. If they are faulty, you may get electrocuted.
- Use appropriate safety systems for work:
- Roofs and fall hazards - use safety harnesses
- Electrically insulated work boots
- Electrically insulated gloves for live working
- Fire retardant clothing
- Arc flash protection equipment for medium and high voltages
- Hard hats as needed
- Sunscreen
- UV rated sunglasses/safety glasses
- Wide brimmed hats for shade
- Drink plenty, hydrate!
- Reflective vests around heavy equipment or traffic
- Competence must be excellent in all areas of work performed
- Attend training as needed
- Perform risk assessments

System Inefficiencies

The system inefficiencies comprise of the following component de-rate factors:

- PV module nameplate DC rating 0.80 - 1.05
- Inverter and Transformer 0.88 – 0.98
- Mismatch 0.97 - 0.995
- Diodes and connections 0.99 – 0.997
- DC wiring 0.97 - 0.99
- AC wiring 0.98 – 0.993
- Soiling 0.30 - 0.995
- System availability 0.00 - 0.995
- Shading 0.00 – 1.00
- Sun-tracking 0.95 – 1.00
- Age 0.70 - 1.00

A well built solar photovoltaic system should be in the vicinity of 77% to 85% overall efficiency conversion from DC to AC at the point of interconnection.

System Size

There are two ways that are traditionally used to quote system sizes

- Installed solar module DC power at STC

- Expected AC output at system interconnection at STC

It is important when designing a system that you quote both values accurately to the customer.

7. Relevant Code Books

Codes in the USA vary by state and you will need to consult with the relevant local authority for the site that you are constructing on to find out which codes apply. Some local jurisdictions have their own codes in addition to these listed below.

If installing in earthquake zones, make sure all equipment meets the relevant earthquake requirements.

Building Codes

- International Building Code (IBC)
- Uniform Building Code (UBC)
- State Specific Building Codes
- Local Building Codes

Electrical Codes

- National Electric Code (NEC) is for residential and commercial systems on the consumer side of the meter. Specifically, section 690 is dedicated to solar photovoltaics.
- National Electric Safety Code (NESC) is for the Utilities. It must be read in conjunction with NEC 690 Solar Photovoltaic Systems to obtain the relevant solar photovoltaic system calculations.

- NFPA 780 Lightning Protection Systems details how to design system lightning protection if needed.

Safety

- OSHA construction standards
- OSHA workplace standards
- OSHA 1910.269 electric power generation, transmission and distribution
- All standards are available at http://www.osha.gov/

8. Relevant Internet Links

- National Renewable Energy Laboratories (NREL)
  - Home Page
    - http://www.nrel.gov/
  - Solar Resource Page
    - http://www.nrel.gov/rredc/solar_resource.html
  - Solar Glossary
    - http://rredc.nrel.gov/solar/glossary/
  - Solar Maps
    - http://www.nrel.gov/gis/solar.html
  - PVWatts calculators
    - http://www.nrel.gov/rredc/pvwatts/
- Department of Energy
  - Home Page
    - http://www.energy.gov/
  - Solar home page
    - http://www.energy.gov/energysources/solar.htm
  - Solar photovoltaics home page
    - http://www1.eere.energy.gov/solar/photovoltaics.html
  - Solar glossary

        - http://www1.eere.energy.gov/solar/solar_glossary.html
- North American Electric Reliability Corporation (NERC)
    - Home page
        - http://www.nerc.com/
- Federal Energy Regulatory Commission (FERC)
    - Home page
        - http://www.ferc.gov/
- Sandia National Laboratory
    - Home Page
        - http://www.sandia.gov/
    - Renewable energy technologies home page
        - http://www.sandia.gov/Renewable_Energy/renewable.htm
    - Solar photovoltaics home page
        - http://photovoltaics.sandia.gov/
- Occupational Health and Safety (OSHA)
    - Home Page
        - http://www.osha.gov/

9. Grid Interconnection Notes

You will need to supply:

- Address of installed system
- System design plans
- Power factor of system generation
- Number of solar modules
- Manufacturer and model type of solar modules
- Installed DC Wp power.
- Installed AC power at the point of interconnection
- Inverter manufacturer(s) and model number(s)
- Sufficient interrupt ratings on breakers and fuses
- For residential and commercial installations only, confirmation of anti-islanding feature.
- Certification of country standards of manufacture of equipment.
- UL 1741 listed or equivalent inverter for residential and commercial installations

For a large utility scale system, you will want the utility to supply the following details:

- Interconnection fault current
- Interconnection voltage

- Transformer configuration
- Relay settings

10. Residential System Notes

A sample residential system is shown on the next page and illustrates the concepts of residential solar photovoltaics. Note that the point of interconnection is at the electrical meter due to the fuse board not being rated for solar photovoltaic (bi-directional power flow) use. We know this because the fuse board had "Line" and "Load" marked on its terminals. Bi-directional fuse boards have no "Line" and "Load" markings on their terminals.

- Make sure that all equipment is de-rated sufficiently for installed conditions.
- Account for all light reflections (increased current) in the electrical design.
- Use appropriate safety systems for work
    - Roof working is common, use fall protection.
    - Risk assessment prior to performing work.
- DC disconnect is mounted at the nearest readily accessible location to solar modules.

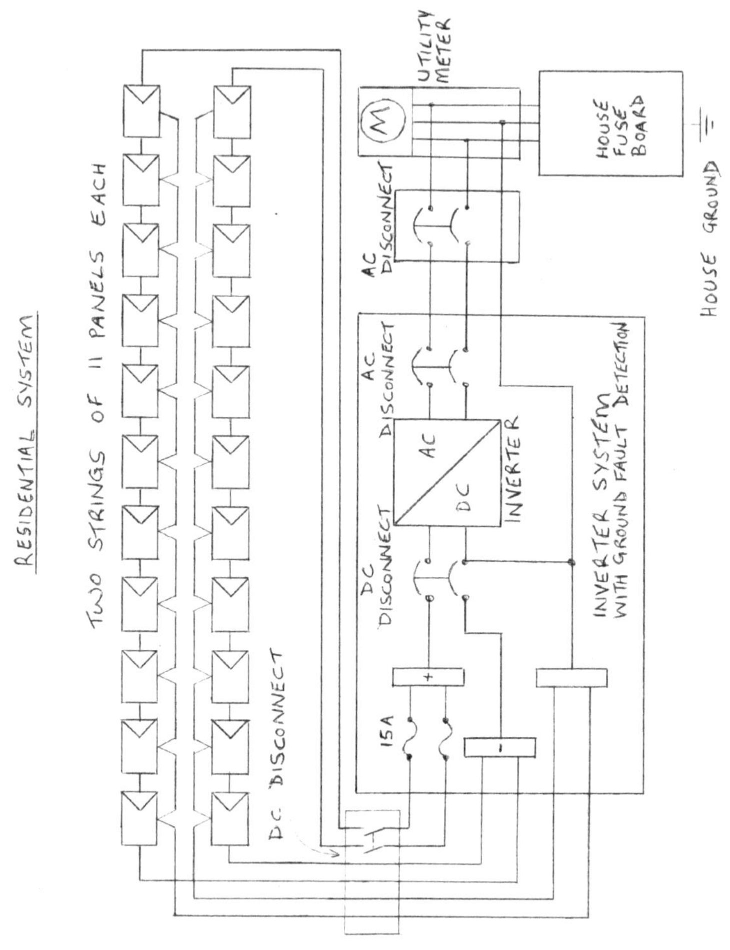
RESIDENTIAL SYSTEM
TWO STRINGS OF 11 PANELS EACH
DC DISCONNECT
15A
DC DISCONNECT
DC
AC
INVERTER
AC DISCONNECT
INVERTER SYSTEM WITH GROUND FAULT DETECTION
AC DISCONNECT
M
UTILITY METER
HOUSE FUSE BOARD
HOUSE GROUND

11. Commercial System Notes

Commercial systems for the purposes of this book is systems from 10 kWp DC to 1 MWp DC at STC. In addition to the requirements of the residential system, follow the points below:

Combiner box standard sizes: 6, 12, 18, 24 fuses

Make sure that the AC electrical system that you feed energy into is rated for back feed between the inverter system and the utility meter. Connect in at the utility meter if not.

The diagram on the following page illustrates the concepts of commercial electrical design for a single 250 kWp DC inverter system.

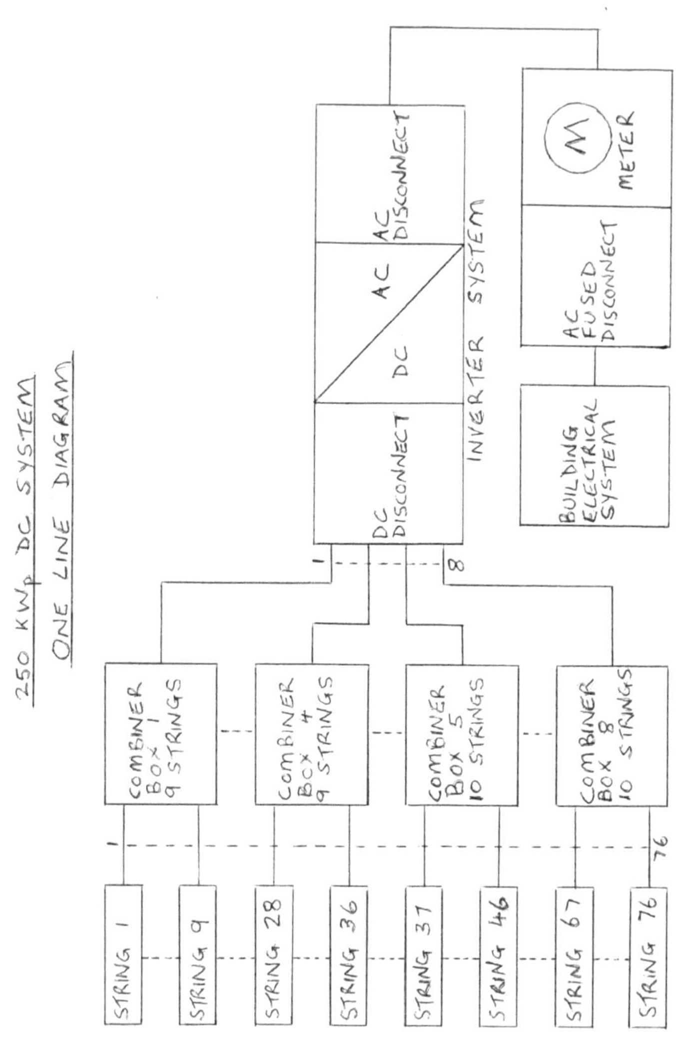
250 KWp DC SYSTEM
ONE LINE DIAGRAM
STRING 1
STRING 9
STRING 28
STRING 36
STRING 37
STRING 46
STRING 67
STRING 76
1
76
COMBINER BOX 1 9 STRINGS
COMBINER BOX 4 9 STRINGS
COMBINER BOX 5 10 STRINGS
COMBINER BOX 8 10 STRINGS
1
8
DC DISCONNECT
DC
AC
AC DISCONNECT
INVERTER SYSTEM
BUILDING ELECTRICAL SYSTEM
AC FUSED DISCONNECT
M
METER

12. Utility System Notes

Utility design for the purposes of this book is systems over 1 MWp DC.

- The utilities are regulated by NERC and FERC.
- All equipment in the AC circuit must be rated for back feed.
- NESC should be used to design the system in conjunction with NEC 690 Solar Photovoltaic Systems.
- General NEC standards do not apply to utilities.
- Pay very close attention to both DC and AC interrupt currents in these large designs.
- Harmonics should be less than 3% at point of interconnection
- Inverters should have ride through capability
- Inverters should have power factor control
- Utility system should be able to handle large power swings from the solar photovoltaic system
- DC systems may be operating at up to 1,000 volts.

A sample design showing the concepts for a 10MWp DC solar photovoltaic utility system is shown on the following pages.

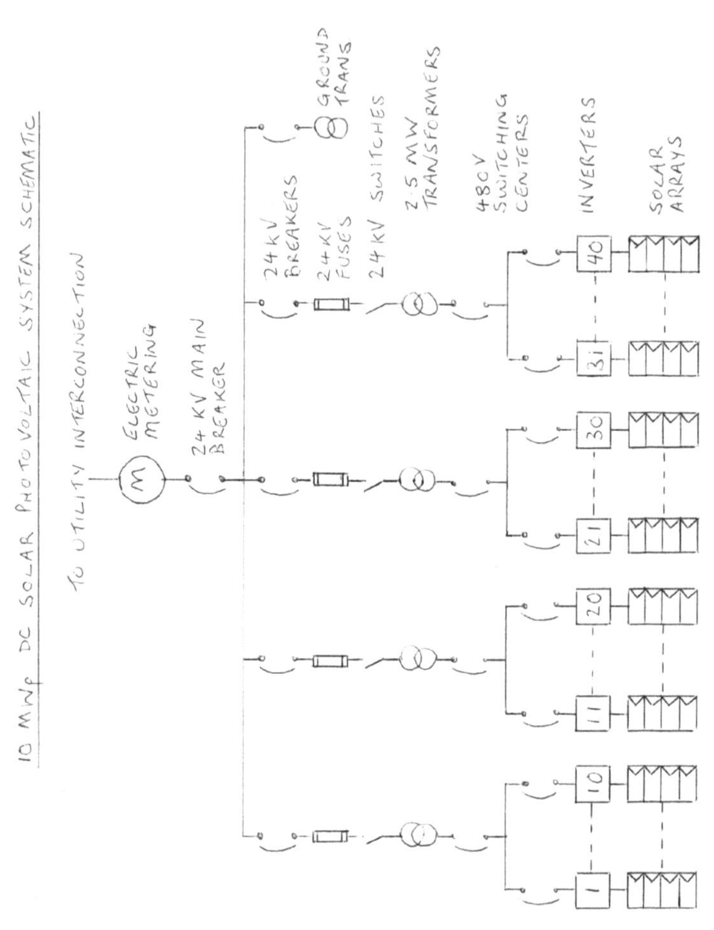
10 MWp DC SOLAR PHOTOVOLTAIC SYSTEM SCHEMATIC
TO UTILITY INTERCONNECTION
M
ELECTRIC METERING
24 KV MAIN BREAKER
24 KV BREAKERS
GROUND TRANS
24 KV FUSES
24 KV SWITCHES
2.5 MW TRANSFORMERS
480 V SWITCHING CENTERS
INVERTERS
1
10
11
20
21
30
31
40
SOLAR ARRAYS

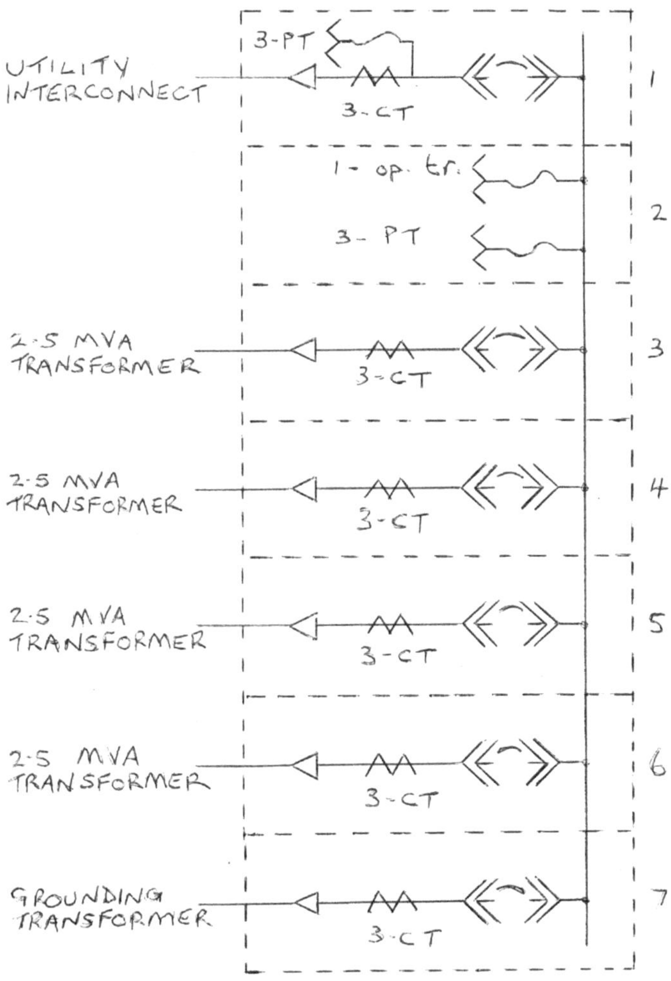
10 MVA DISTRIBUTION
UTILITY INTERCONNECT
3-PT
3-CT
1
1- op. tr.
3- PT
2
2.5 MVA TRANSFORMER
3-CT
3
2.5 MVA TRANSFORMER
3-CT
4
2.5 MVA TRANSFORMER
3-CT
5
2.5 MVA TRANSFORMER
3-CT
6
GROUNDING TRANSFORMER
3-CT
7

13. NREL Annual Insolation Chart

Web location: http://www.nrel.gov/gis/solar.html

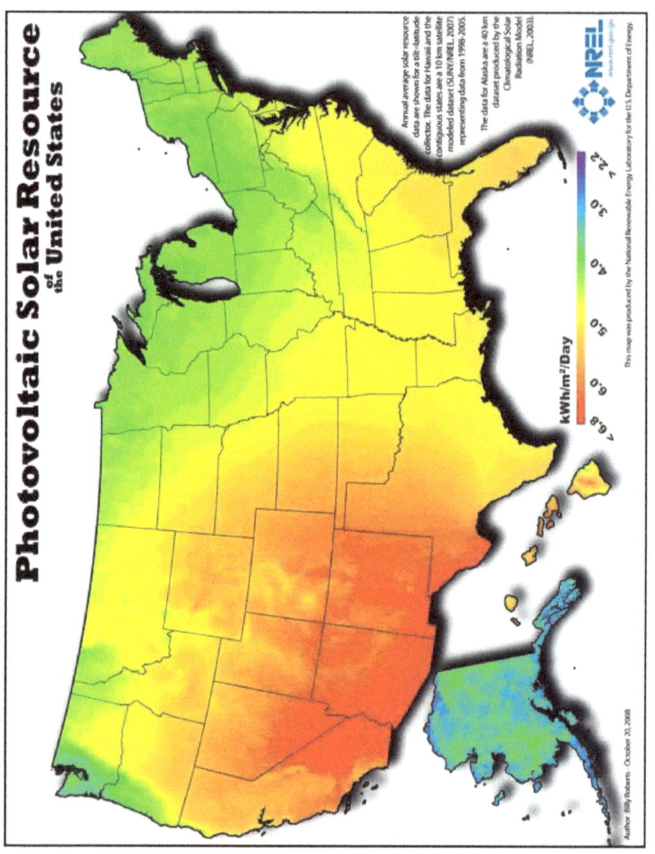

14. NREL Monthly Insolation Charts

Web location: http://www.nrel.gov/gis/solar.html

January

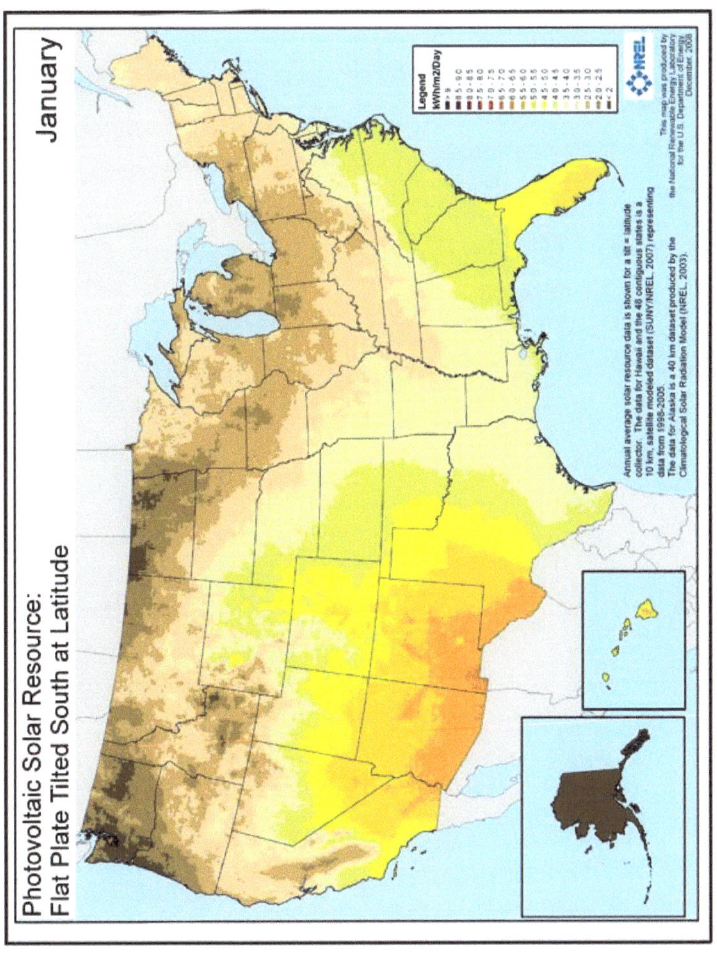

February

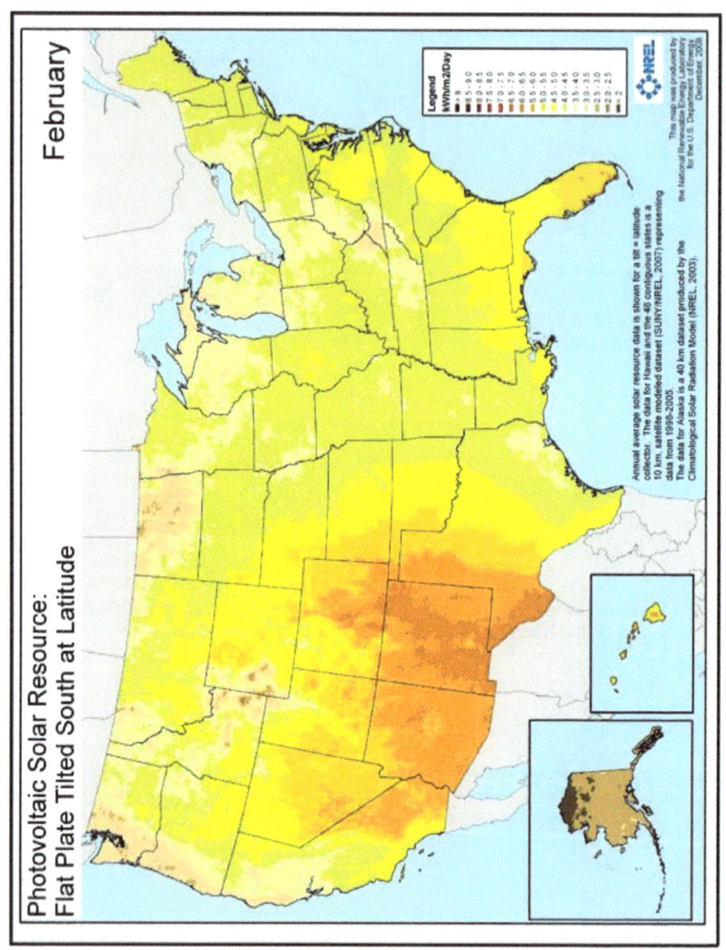
Photovoltaic Solar Resource:
Flat Plate Tilted South at Latitude
February
Legend
kWh/m2/Day
NREL
Annual average solar resource data is shown for a tilt = latitude collector. The data for Hawaii and the 48 contiguous states is a 10 km, satellite modeled dataset (SUNY/NREL, 2007) representing data from 1998-2005.
The data for Alaska is a 40 km dataset produced by the Climatological Solar Radiation Model (NREL, 2003).
This map was produced by the National Renewable Energy Laboratory for the U.S. Department of Energy December 2008

Solar Photovoltaic Resource by Steven Magee

<u>March</u>

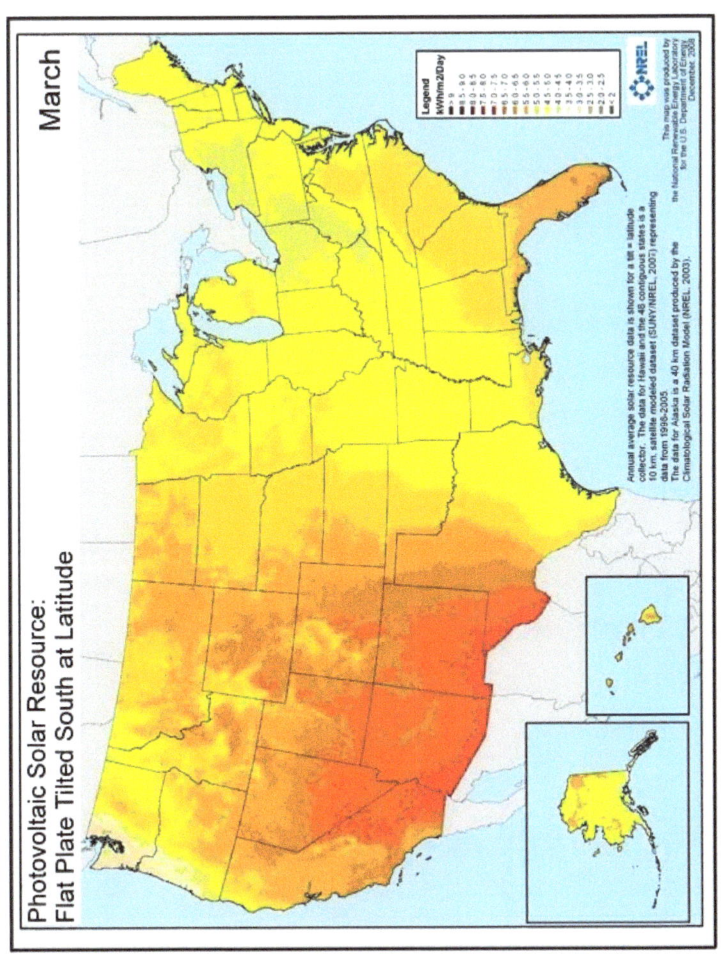

April

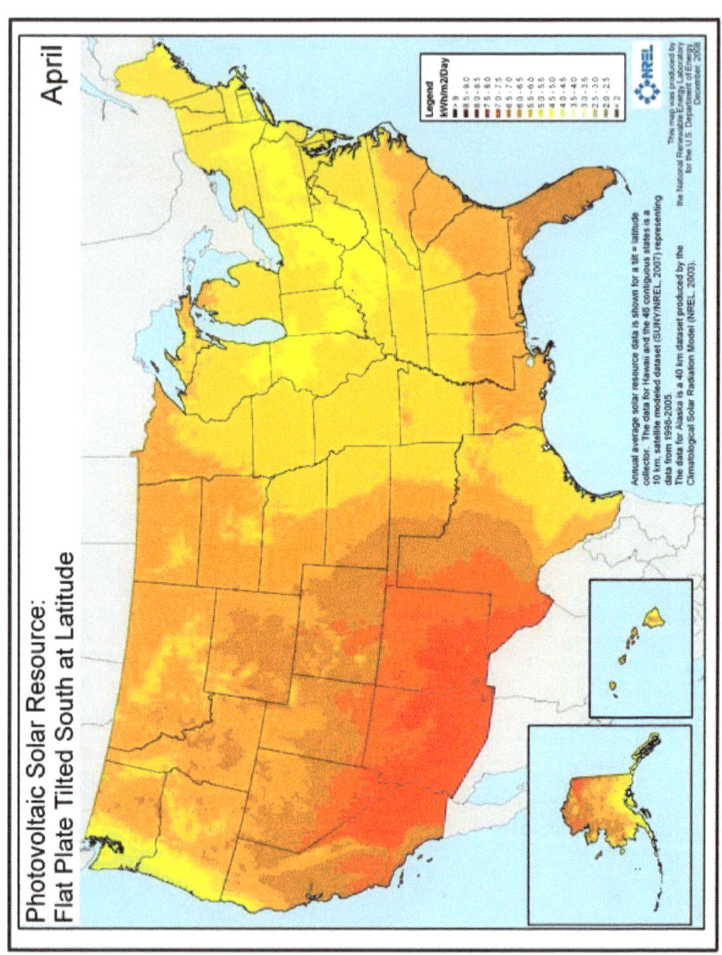
Photovoltaic Solar Resource:
Flat Plate Tilted South at Latitude
April

Solar Photovoltaic Resource by Steven Magee

<u>May</u>

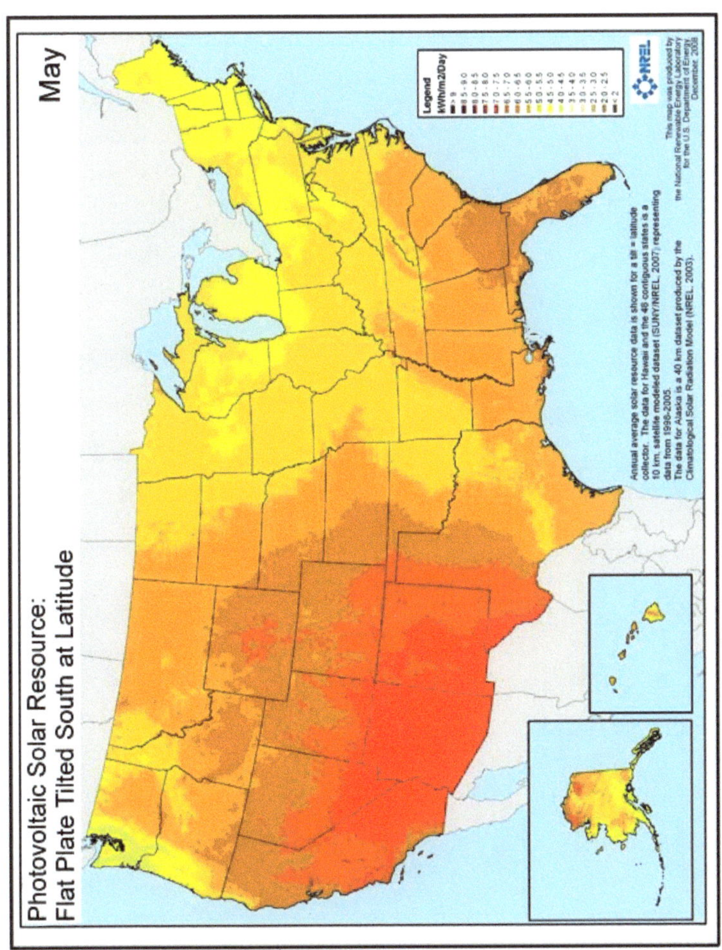

June

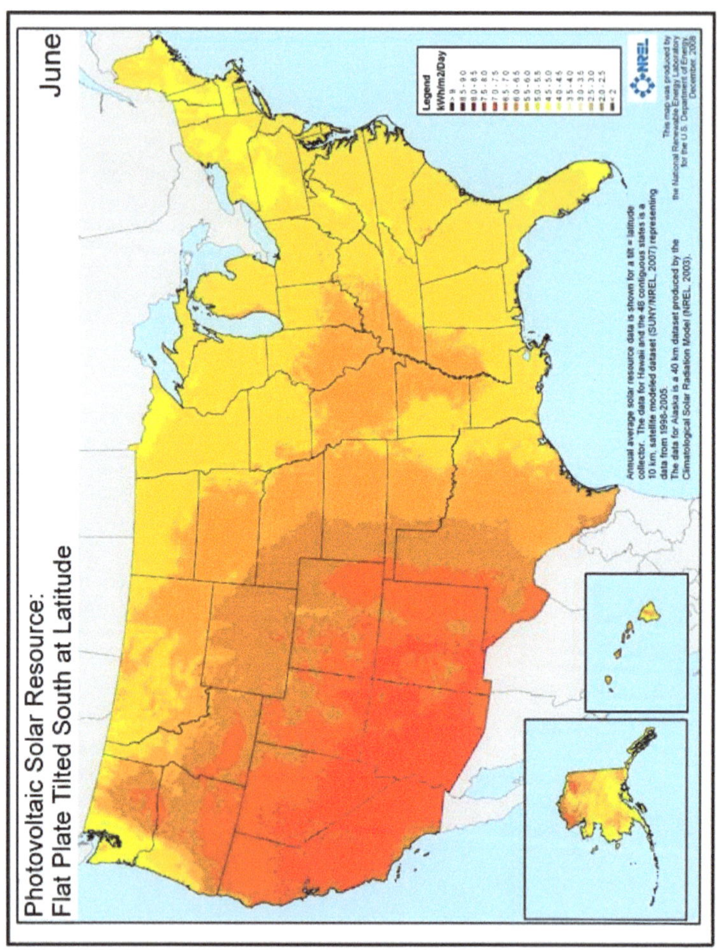
Photovoltaic Solar Resource:
Flat Plate Tilted South at Latitude
June
Legend
kWh/m2/Day
Annual average solar resource data is shown for a tilt = latitude collector. The data for Hawaii and the 48 contiguous states is a 10 km, satellite modeled dataset (SUNY/NREL, 2007) representing data from 1998-2005.
The data for Alaska is a 40 km dataset produced by the Climatological Solar Radiation Model (NREL, 2003).
NREL
This map was produced by the National Renewable Energy Laboratory for the U.S. Department of Energy
December 2008

July

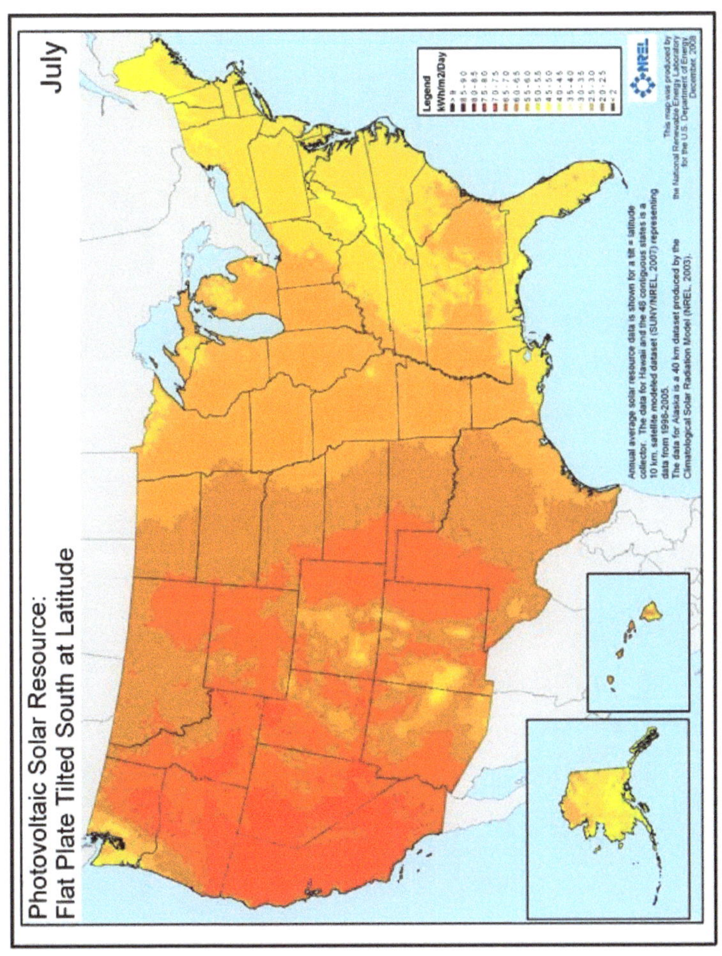
Photovoltaic Solar Resource:
Flat Plate Tilted South at Latitude
July
Legend
kWh/m2/Day
NREL
Annual average solar resource data is shown for a tilt = latitude collector. The data for Hawaii and the 48 contiguous states is a 10 km, satellite modeled dataset (SUNY/NREL, 2007) representing data from 1998-2005.
The data for Alaska is a 40 km dataset produced by the Climatological Solar Radiation Model (NREL, 2003).
This map was produced by the National Renewable Energy Laboratory for the U.S. Department of Energy December, 2008

<u>August</u>

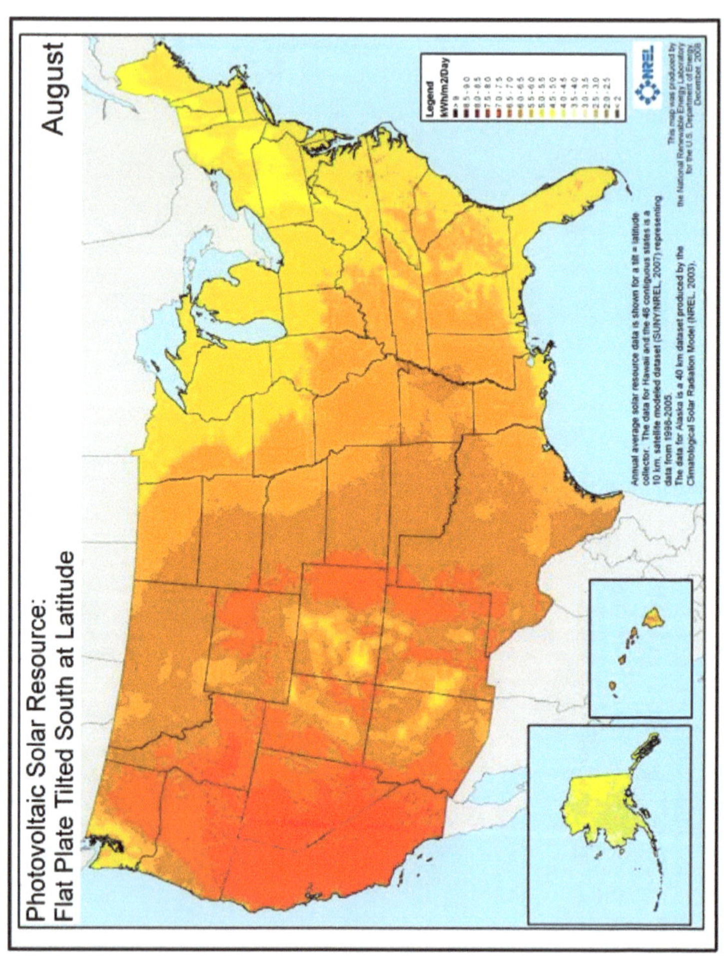
Photovoltaic Solar Resource:
Flat Plate Tilted South at Latitude
August
Legend
kWh/m2/Day
Annual average solar resource data is shown for a tilt = latitude collector. The data for Hawaii and the 48 contiguous states is a 10 km, satellite modeled dataset (SUNY/NREL, 2007) representing data from 1998-2005.
The data for Alaska is a 40 km dataset produced by the Climatological Solar Radiation Model (NREL, 2003).
NREL
This map was produced by the National Renewable Energy Laboratory for the U.S. Department of Energy. December 2008

Solar Photovoltaic Resource by Steven Magee

September

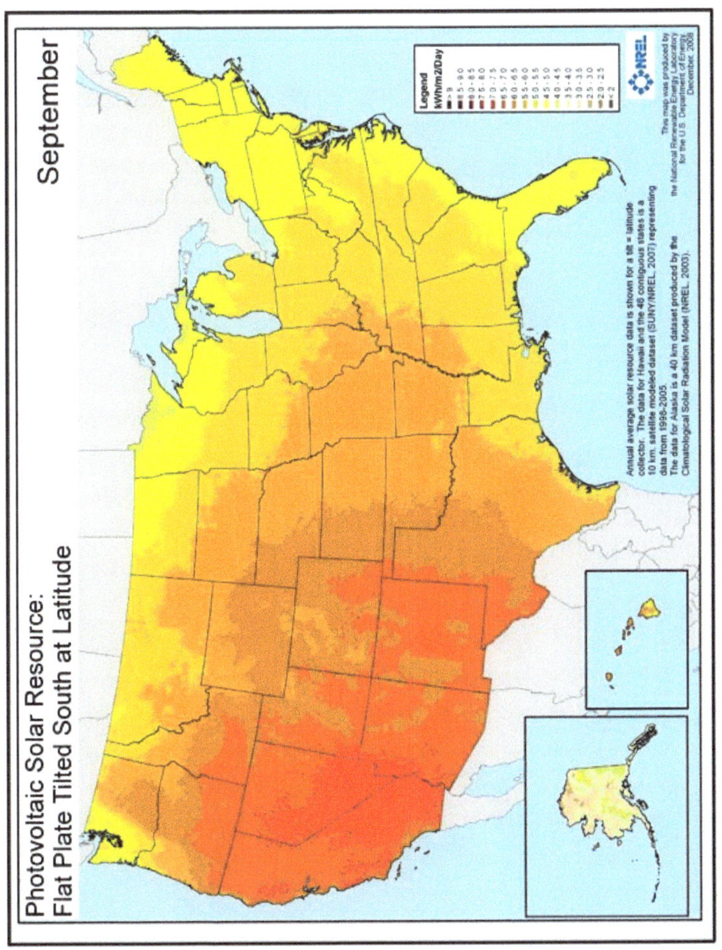

October

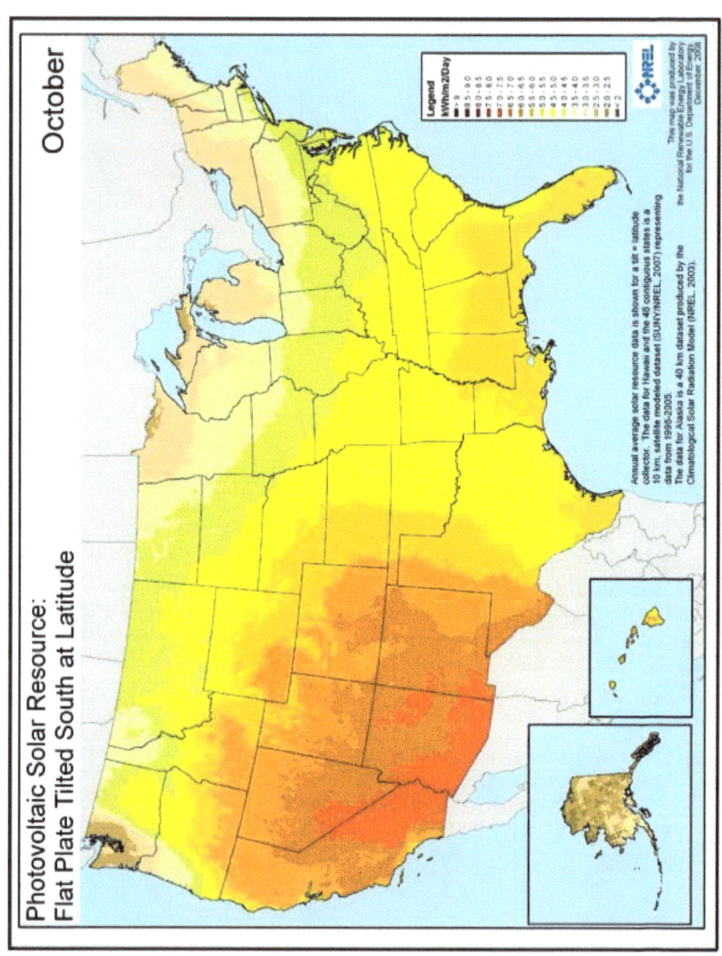
Photovoltaic Solar Resource:
Flat Plate Tilted South at Latitude
October

Solar Photovoltaic Resource by Steven Magee

November

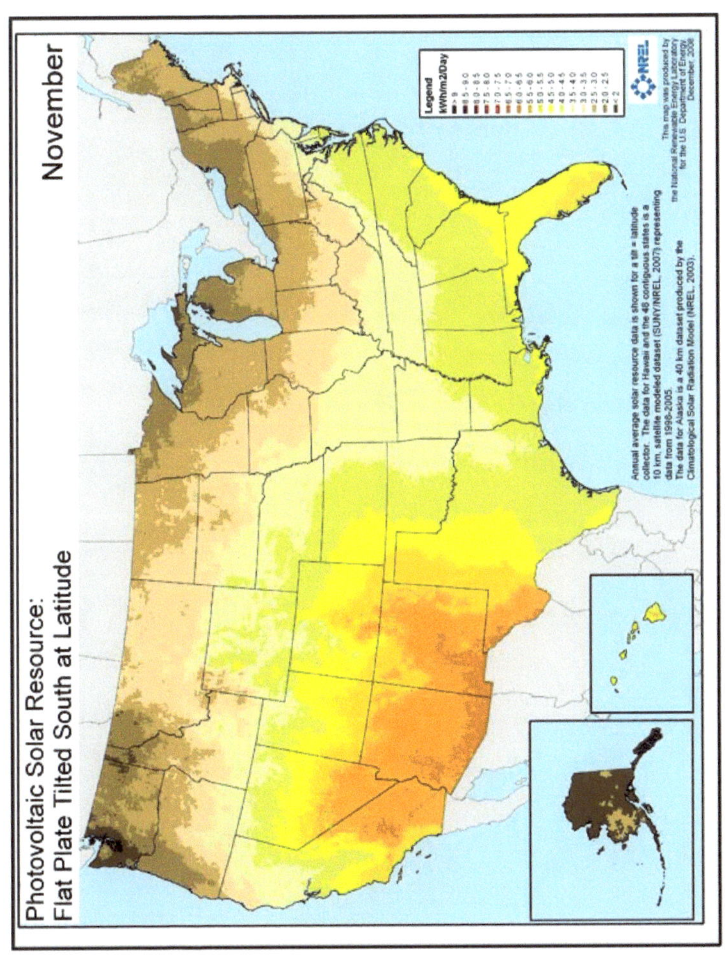

December

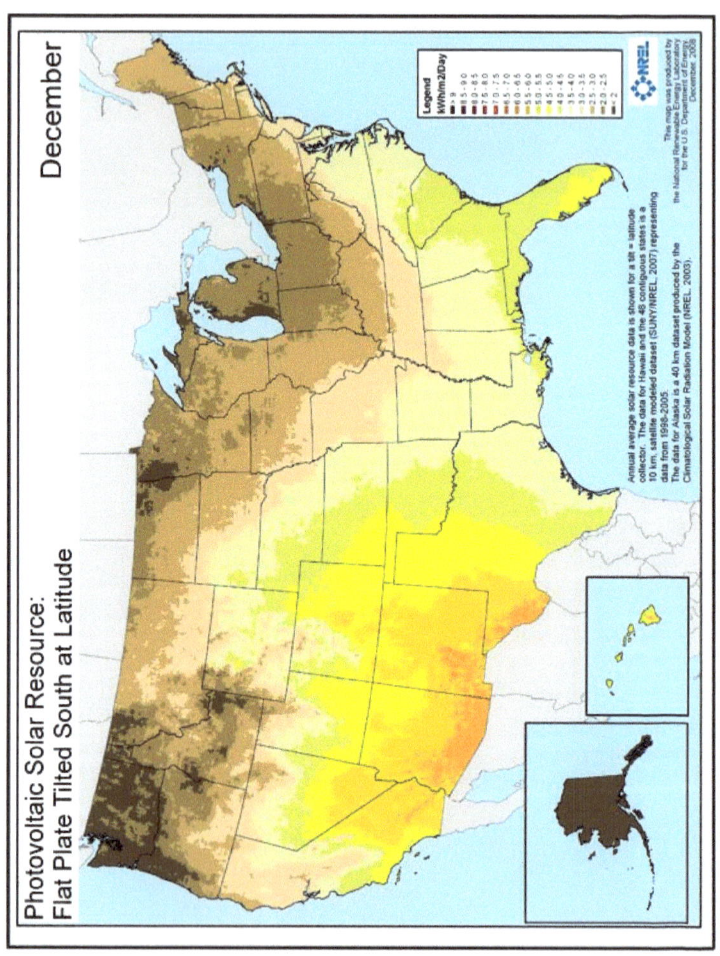
Photovoltaic Solar Resource:
Flat Plate Tilted South at Latitude
December
Legend
kWh/m2/Day
Annual average solar resource data is shown for a tilt = latitude collector. The data for Hawaii and the 48 contiguous states is a 10 km, satellite modeled dataset (SUNY/NREL, 2007) representing data from 1998-2005.
The data for Alaska is a 40 km dataset produced by the Climatological Solar Radiation Model (NREL, 2003).
This map was produced by the National Renewable Energy Laboratory for the U.S. Department of Energy. December 2008
NREL

15. System Sizing Notes

DC to AC conversion derating factor is 0.77 generally.

Single axis tracker increased annual energy from fixed tilt is 1.24 generally.

Dual axis tracker increased annual energy from fixed tilt is 1.3 generally.

System aging is an average of 1% decrease per year from new, or approximately a 20% decrease after twenty years.

Temperature adjustments for solar photovoltaic module power are approximately -0.485%/°C for silicon

Irradiance and system current are approximately proportional.

PVWatt calculators at the NREL website:

http://www.nrel.gov/rredc/pvwatts/ to size solar photovoltaic systems.

Consult with the manufacturers data sheets for actual values for the system that you are working with. You must remember that solar photovoltaics is more of an art than a science and these figures are approximations. It is common for some systems to over perform and others to under perform. It depends on local conditions.

16. Solar Module Notes

All solar modules should be tested to the Underwriters Laboratory (UL) 1703 standard or equivalent in the USA.

The main solar technologies in use currently are:

- Mono-crystalline silicon wafers (also called single crystal)
- Poly-crystalline silicon wafers (also called multi-crystalline)
- Amorphous silicon thin film
- Cadmium telluride thin film
- CIGS thin film

Be careful when using manufacturers data sheets for their solar modules. The module power value generally has percentage tolerances for negative and positive adjustments to this figure. This is a reflection of the imprecision of the solar photovoltaic cell manufacturing processes. Due to this it is common for individual strings of solar modules to either over or under perform in the system. For example, the cloud effect combined with a 10% high performing string of solar modules could produce 165% more than the rated STC current from the string. If there are other reflections present then it will produce even more current! If your system under performs it may be due to having the bulk of the supplied modules performing at lower than normal power ratings. At least one major manufacturer does not list any negative adjustments on its solar photovoltaic module data sheets, only positive adjustments. This greatly helps with meeting the system performance requirements. It would be good to see other manufacturers follow their lead.

Electrically, a solar photovoltaic module is very simple device. It turns light into current at a rate that is proportional to the irradiance it receives. It does this at a voltage that is dependent on the temperature of the solar cells. The solar module will keep indefinitely producing current proportional to the levels of irradiance it receives, even if these levels are much higher than STC due to reflections and lensing. The solar module does this conversion in real time, there is no delay in the conversion process.

A simple approximation of a solar module is that it is a current limited source and that the upper limit for current is entirely dependent on the amount of light received by it!

Due to this, it pays to use the maximum fuse size as recommended by the manufacturer of the solar module.

If the irradiance levels are very low at either dusk or dawn then the voltage will be somewhere between 0V and Voc when adjusted for solar cell temperature. At night time it will be at or very close to 0V. The inverter system that the solar photovoltaic modules are connected to knows when to turn on and turn off by monitoring the voltage rise at dawn and the voltage fall at dusk. The inverter system will generally switch on and off at preset voltage levels that have been entered into it during installation.

Voltage imbalances between the strings may occur at dawn and dusk if the solar modules are facing in different directions from each other on the same inverter system.

Solar modules commonly have what is known as "bypass diodes" installed in them. These are installed for shading effects and for bypassing electrically faulty solar cells inside the solar photovoltaic module. This enables the other modules in the string to function if one is shaded of faulty.

Less common on grid tied systems are "blocking" diodes. One per string can be installed and this prevents reverse currents from flowing into the strings from the system. The string fuses now protect against this by blowing if the reverse current gets too high in a grid tied solar photovoltaic system. It is for this reason that the manufacturer limits the maximum size of the string fuse. If the fuse is above the maximum size recommended and reverse currents occur, then the solar modules in the string may be damaged beyond repair. Never exceed the maximum recommended string fuse size.

Solar photovoltaic modules can blow the string fuse for any of the following reasons:

- Producing too much current (irradiance and reflections)
- Reverse currents (voltage imbalance between strings)
- Ground faults
- Short circuits

If you take a look at the solar module data sheets from the various manufacturers you will see graphs showing the above concepts for their products. Take time to look at the data sheets and understand them, they have a lot of information in them. Don't build a system with their product until you have read and fully understood the data sheet for it.

The pictures on the following pages demonstrates the concepts of solar modules.

This picture shows the large photo-diodes of a mono-crystalline silicon solar photovoltaic module as seen through the glass.

This picture shows the bypass diodes that most solar photovoltaic modules have installed inside their solar photovoltaic module junction boxes.

This picture shows the internal electrical diagram of a silicon solar photovoltaic module. As you can see, it is a collection of very large photo diodes wired together into strings. The bypass diodes electrically bypass the internal strings of solar cells if they are either faulty or shaded.

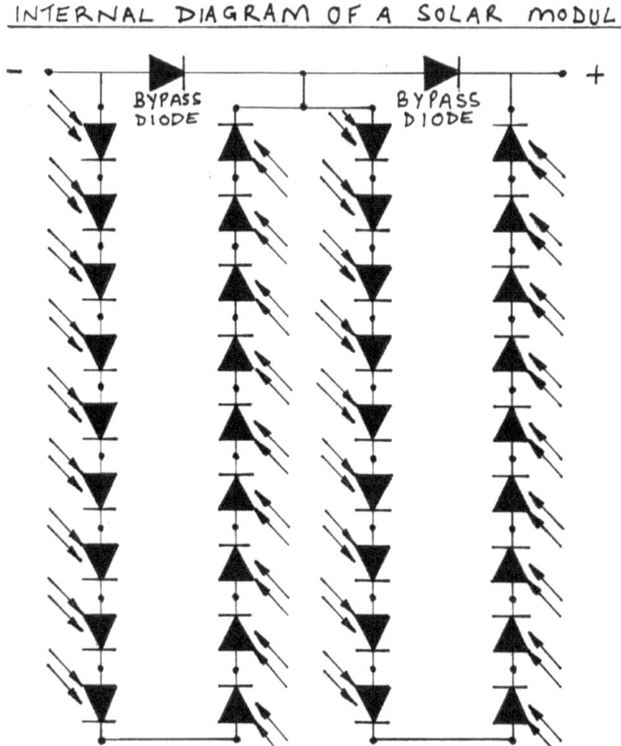

A SILICON SOLAR MODULE CONSISTS OF LARGE PHOTO DIODE CELLS AND BYPASS DIODES

Solar modules are first wired in series, this is called "stringing", to increase their DC voltage output. Then many strings are connected in parallel to increase the current output of the solar system. Raising the voltage and current is needed for the inverter system to work properly and to increase the efficiency of the system.

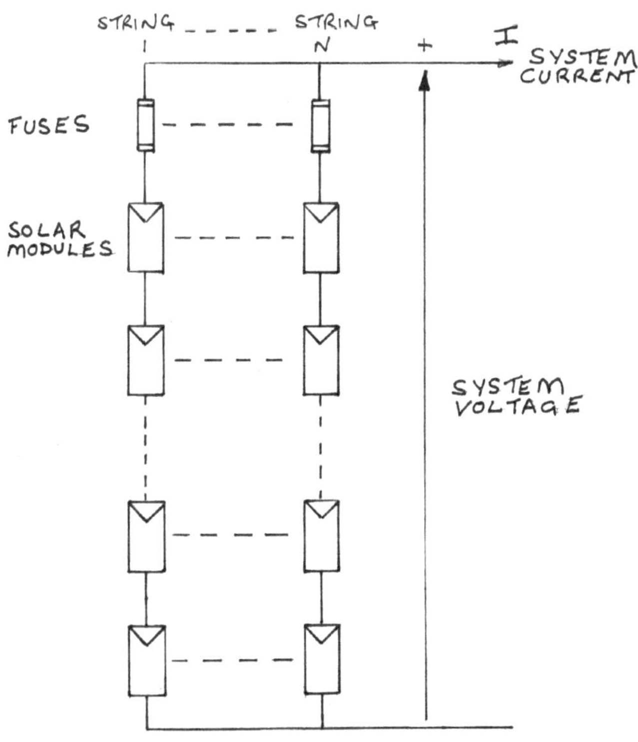

For residential installations, the supporting infrastructure costs vary little between the different technology types for installation, operation and maintenance.

This dynamic starts to change as the system gets larger with commercial and utility installations. Just an decrease in conversion efficiency of 1% can add acres of land use and supporting infrastructure to a 10 MWp project. The industry is constantly trying to find more efficiency out of their products due to this.

If land use is a significant project expense or is limited, then you probably should be using the more efficient technologies.

Some areas have very cheap land and if this is the case, the more important figures to look at are the supporting infrastructure costs for a larger installation and the ongoing annual operation and maintenance costs for a larger power generation facility. In the dry deserts of the South West USA, these operation and maintenance costs are generally quite low.

Thin film has better performance in low light conditions when compared to the silicon wafer technologies and this may be a factor to consider if the installation area is frequently cloudy.

If the failure rate is the same across all technologies, then the you will be replacing more equipment during the year on a less efficient installation during routine maintenance of the system. Manufacturers appear to be reluctant to release solar module failure rates for their solar module products and on any project, I have become accustomed to assuming a 1% annual failure rate in absence of this information. Unusual and unexpected severe weather conditions may increase this failure rate. Severe weather insurance is recommended to cover these events.

Operation and maintenance costs will be costly for some projects due to:

- Poor site selection
- Poor equipment selection
- Poor wildlife assessment
- Poor build quality

Performing due diligence at the start of any project will keep these problems under control. Ensuring that a highly skilled solar photovoltaic consultant is overseeing large projects from conception through to design and onto construction and commissioning will help to prevent these problems from occurring.

Site selection is probably the most important part of any large solar photovoltaic system project and can be the determining factor in a projects overall success. On any large project it is recommended to install a small version of the proposed system for the first year prior to constructing the main project so that an assessment can be made of how the system will perform in that location and design changes can be incorporated into the main project once all the risks are known. In the large scale solar photovoltaics field it is the tortoise that is generally more successful, try not to be the hare.

Be wary of locations close to the ocean and take the necessary precautions with equipment selection so that high winds, salt spray and corrosion don't become a problem. Salt will deposit itself onto the surface of the modules and if it is excessive, you will be frequently washing the solar photovoltaic modules. Expect the operation and maintenance costs to be higher when installing next to the ocean.

Frame-less solar photovoltaic modules are becoming available and these have better self cleaning properties during rains when compared to framed modules. Frame-less modules are not as strong as framed modules and you will need to assess if this is a problem for your installation or not. Some frame-less modules do not require a ground connection and this can make a significant saving on a large project.

The table on the next page shows the different technologies.

Module Type	Module Wattage	Efficiency	Modules in Strings	# Modules	10 MW Area	Combiner Boxes
Mono	235W	14.42%	14	42,560	100 acres	254
Poly	215W	14.48%	16	46,512	99.74 acres	243
a-Si	135W	9.49%	2	74,076	151.8 acres	3,087
CdTe	70W	9.722%	6	142,858	148.3 acres	1,985
CIGS	150W	7.61%	5	66,667	184.5 acres	1,112

17. Inverter Notes

All residential and commercial solar photovoltaic grid tie inverters should comply with UL1741 in the USA. This ensures reliability throughout the various climate zones that the USA has to offer.

AC solar modules are starting to appear and this is at the smallest end of the inverter spectrum. Each solar module has it's own inverter attached to it, commonly called a "micro" inverter. This gives great MPPT performance. This is the ideal configuration for systems that may suffer from:

- Shading
- Have solar modules facing in different directions from each other
- Heavy dirt accumulation

Micro inverters can be useful where there is nowhere to mount the traditional inverter on a domestic installation, or if the owner considers an inverter box ugly. The micro inverters can be attached to the mounting system and hidden from view.

When connecting strings to the inverter, do not mix and match solar module types and connect them to a single inverter, this is a recipe for disaster. If you desire to install the different technology types on the same site, then each solar module technology must have its own inverter system dedicated to it.

Module mismatch in strings is also an issue for the inverter MPPT. Do not connect different wattage modules together in the same string as the higher wattage modules will try and push their current through the lower wattage ones. This will waste power. Each individual solar string needs to have identical wattage modules in it.

Foreign manufactured inverter systems may arrive with the inverter set for a different voltage or frequency of AC electricity. Always check that the inverter system matches the grid system that you are connecting into.

Transformer-less inverters are on the market and these require a different set up on the DC solar field. Basically half the solar DC field is set to positive polarity and the other half of the field is set to negative polarity. These are a relatively new to market inverter system and have not established themselves with a long history yet. They have very high DC to AC conversion efficiencies.

18. Switchgear Notes

Older AC switchgear was not constructed with solar power generation in mind and may only be rated for electricity flow in one direction. If so, they will have markings on the electrical terminals that say either "load" or "line".

All switchgear in the AC circuit on a grid connected solar photovoltaic power system will need to be rated for bi-directional current flow. Due to their bi-directional operation, these electrical systems have no "load" or "line" markings on the electrical terminals.

The AC switchgear that needs to have this bi-directional rating should only be that equipment that is between the solar photovoltaic inverter system and the point of interconnection to the utility. In a residential and commercial system, this point of interconnection to the utility is at the utility meter. In other words, only the switchgear that energy reversal takes place in needs to be rated for bi-directional operation.

19. Distribution Notes

If you are working at this level, it is extremely important that you are competent with working with this type of equipment. The voltages at this level are very unforgiving and death can easily result from misguided activities with this equipment. People die every year by electrocution on these systems, it is an occupational hazard. OSHA standard 1910.269 for electric power generation, transmission and distribution details the safety requirements for this equipment for the utilities. Work safe and be safe.

Solar photovoltaic systems feature a bi-directional current flow that reverses twice per day. During the day time the solar power system will generate energy into the grid and during the night, the energy will reverse and it will consume a small amount of power from the grid. As such, all distribution equipment in a grid connected solar photovoltaic power system will need to be rated for bi-directional current flow.

Distribution generally covers the following items

- Transformers
- Distribution poles
- Medium voltage switchgear
- Relays
- Harmonics

Transformers

Dry type transformers are generally used in solar photovoltaics and have very high efficiencies of over 99% conversion. Some people are still using oil filled transformers and the operation and maintenance costs are higher with these due to needing more attention and care.

Distribution Poles

Many installations use distribution poles to connect into the point of interconnection (POI) at the grid. It is a cheaper way of distribution when long distances are involved. For short distances of less than about 1 mile there really isn't any difference between using distribution poles or going underground, if the underground route is easy to put ducts and conduits into.

Medium Voltage Switchgear

Medium voltage switchgear is generally of the type:

- Gas insulated
- Vacuum insulated

The choice of which type to use is generally driven by costs. The higher power switchgear is generally vacuum insulated.

Protective Relays

Protective relays control the tripping mechanisms on the switchgear. They are set up according to the desired fault levels that the designer has calculated for each switch on the system.

Harmonics

Inverter systems cause harmonics. Normally, total harmonic distortion (THD) should be below 3% and is quite acceptable to feed power into the grid at these low levels.

20. De-rating Notes

De-rating is very, very important in the solar photovoltaics field. Outdoor equipment is operating in hot temperatures and some of it it faces the sun all day long. Solar modules have current flowing in them and they dissipate heat through electrical losses, so they get even hotter than ambient temperatures.

The same is true for all electrical equipment. All electrical equipment dissipates heat when current is flowing through it. Operating this equipment in conjunction with the outdoor high ambient summertime temperatures and currents stresses it, so it is very important that you de-rate it more than sufficiently for the expected installed conditions.

De-rating is most important and must be applied to each individual component of the system for reliability:

- De-rate for highest annual ambient temperatures
- De-rate for highest expected internal enclosure temperatures
- De-rate for highest expected cable temperatures
- De-rate for highest possible currents when accounting for ALL sources of system light reflections

When in doubt, de-rate!

21. Ancillary Systems

These systems are not essential and are generally installed on larger sites, particularly if they are unmanned:

- Weather Station
- Security System
- Video Surveillance
- Metering
- Inverter Monitoring
- String Monitoring
- Monitoring Services

Weather Station

Many solar sites install weather systems, particularly if they are unmanned. They can provide useful information that can be used to verify the power output of the system. Most solar site weather systems include the following sensors:

- Reference solar cell with temperature
- Irradiance sensor
- Solar module temperature sensor
- Ambient temperature
- Humidity
- Wind speed and direction

- Rain gauge

Lets look at each of these in detail:

Reference Solar Cell With Temperature

This reference cell will be made from the same solar cell type as the installed solar modules contain. It is a factory calibrated solar cell. It has a temperature probe attached to it to record the cell temperature. This allows the cell reading to be adjusted to the correct value.

Irradiance Sensor

This is used to record the irradiance levels and is sometimes referred to as a pyranometer.

Solar Module Temperature Sensor

The solar module temperature sensor is attached to the rear surface of one of the installed solar modules. This allows the operating temperature of the solar module to be known and for the solar module output to be adjusted for temperature effects.

Ambient Temperature

The site ambient temperature is recorded for historical purposes.

Humidity

The site humidity is recorded for historical purposes.

Wind Speed and Direction

The wind speed and direction can be used to assess the cooling effects from this on the solar modules.

High wind speeds are a concern for system damage from flying debris. After a high wind speed event has been detected, a site visit should be arranged to check on the equipment.

Rain Gauge

It is useful to know the historical rainfall, so that you can schedule cleaning. If you have a site with consistent rainfall, then it is unlikely that you will need to clean the solar modules regularly. If you have a site that only gets sustained rainfall on one or two occasions per year, you may want to schedule solar module cleaning more frequently.

If any part of the weather system fails, then it is usually replaced. Weather system parts are generally regarded as consumables due to the harsh environment that they work in.

Security System

On remote sites a security system is usually installed. The security system will usually use beam break sensors within the caged areas of the solar site. Keep these beam break sensors clean to ensure that false alarms do not occur.

Thefts have not been a significant problem in the industry yet. It helps that the solar modules are firmly attached to their mounting structure and that tools are needed to remove them. Generally most thieves are after the copper wiring. This is why it is important to always check the grounding systems of your installations, as they may have been stolen.

As public understanding of solar photovoltaic power systems increases and demand for the solar modules goes mainstream, this may develop. It is a good idea to install the security system during construction of the project to keep your insurance premiums down on the system.

Video Surveillance

On remote sites usually internet based video surveillance cameras are installed with the pan, tilt and zoom (PTZ) feature. These will be low light level cameras that can work under moonlight. They are mounted in weather resistant housings.

Metering

The electrical meter for the solar photovoltaic system can be connected to the internet. This is useful for billing purposes if the energy is being sold to a customer. It also is very useful for detecting a problem at a remote site. If the site is not producing energy, then it can be quickly detected by monitoring the meter.

Inverter Monitoring

Most inverters have the capability to be monitored and controlled. This is useful for power factor control and for detecting inverter faults. If the inverter has power factor control, then it can be used to provide power factor control to the grid that it is connected to. The utilities like to have this feature if it is available.

The inverters have many fault codes and it is useful for a remote site to know the fault code in order to schedule a repair visit.

A nice feature that the inverters have is the ability to monitor the DC inputs to them. This is commonly referred to as "zone monitoring". When monitoring at the zone level, it is easier to detect problems with the solar module strings by comparing the zones to each other. If one zone is consistently reading less than another equally sized zone, then you will need to schedule a repair visit to that zone.

String Monitoring

Monitoring can be done at the string level. On a residential system this is normal for the inverter to be monitoring at this level due to the small number of strings connected to it. On commercial and utility systems, this would be installed in the combiner boxes.

On utility systems with several thousand solar photovoltaic strings, this is a lot of monitoring and as such, it is also expensive to install and maintain. It is nice to have, but not really needed. Good inverter zone monitoring should be able to detect an individual string failure.

Monitoring Services

There are many companies now offering monitoring services for solar power generation systems. While not so useful for residential and commercial applications where there will be people present to check on the system, they are extremely useful for remote sites.

The monitoring center will send out automatic email notifications to you when problems are detected on site. This enables a response to be arranged quickly before too much energy generation is lost. On a large solar photovoltaic power generation system, it doesn't take much time of the system being out of service before thousands of dollars of energy generation has been lost.

These monitoring services are invaluable on large generation systems and their expense is far outweighed by the revenue saved by keeping the solar power generation system in full service.

22. Power Purchase Agreements (PPA)

Many solar installations are built using power purchase agreements(PPA) that are used to finance the project. A PPA is quite a simple idea. A loan to finance the project is taken out by the company that builds it. It is built on land that is owned by the customer. Once in operation and producing energy, that energy is sold to the customer who has agreed to purchase all energy produced by the system over twenty years. The sales of energy during the twenty years are used to pay for the operation and maintenance costs of the system and the original loan interest and principal.

After about ten years the loan is repayed in full and all future sales of energy produce profit at that point after routine costs. After the twenty years is up, the customer can either choose to enter into another PPA for energy supplied, have the system removed from their property or purchase the system.

The PPA is a great concept and when a system performs as planned or even better, they are a great way of building solar photovoltaic systems. Due diligence at the planning stage ensures this.

23. Site Licensing Agreements (SLA)

Site licensing agreements (SLA) are the partner to the power purchase agreement (PPA). If you are building on a piece of property that is owned by someone else, you will need one of these. It is basically a contract for land use for the duration of the project.

It is important that the SLA is well written and incorporates the following:

- Term of land use
- Unrestricted access for the duration of the project
- No impacts to the solar photovoltaic power generation system from future surrounding developments (generally shading & dirt problems)
- Energy costs
- Guaranteed power purchasing by the customer
- Customer maintains the point of AC interconnection in service at all times
- Customer pays compensation if the point of interconnection is not available for power production.
- End of lease needs to be well defined for the options to enter into another PPA, purchase the system, or have the system removed.

24. Solar Photovoltaic Lexicon

- EPC - Engineer, Procure and Construct
- Impp - Current at maximum power point operation
- Insolation - Time based measurement of solar irradiance. Units are watts per square meter per day
- Irradiance - Solar radiation power level. Units are watts per square meter
- Isc - Current at short circuit operation
- kW - Kilowatt (1,000 watts)
- kWh - Kilowatt Hour (1,000 watt hours)
- m^2 - Square Meter
- MPP - Maximum power point of the optimum DC current and voltage values to produce peak power.
- MPPT - Maximum power point tracking, the inverter does this automatically to keep the DC system producing peak power.
- MW - Megawatt (1,000,000 Watts)
- MWh - Megagwatt Hour (1,000,000 watt hours)
- Net zero - The solar photovoltaic system generates the same annual energy as is consumed annually by the residential or commercial premises where it is installed.
- PPA - Power purchase agreement
- SAT - Single axis tracker
- SLA - Site licensing agreement

- STC - Standard Test Conditions
- UL - Underwriters Laboratory
- UL1703 - Standard for solar module testing
- UL1741 - Standard for inverter testing
- Vmpp - Voltage at maximum power point operation
- Voc - Voltage at open circuit operation
- Watt - Unit measure of electrical power
- W/m^2 - Watts per square meter
- $W/m^{2/}$Day - Watts per square meter per day
- Wp - DC solar power at Standard Test Conditions

25. References

- NFPA National Electrical Code (NEC)
- IEEE National Electric Safety Code (NESC)
- International Building Code (IBC)
- Uniform Building Code (UBC)
- Occupational Safety and Health Administration www.osha.gov
- National Renewable Energy Laboratories www.nrel.gov
- Sandia National Laboratory http://www.sandia.gov/
- United States Department of Energy www.energy.gov
- North American Electric Reliability Corporation (NERC) http://www.nerc.com/
- Federal Energy Regulatory Commission http://www.ferc.gov/
- Solar Photovoltaics for Consumers, Utilities and Investors by Steven Magee
- Solar Photovoltaic Training for Residential, Commercial and Utility Systems by Steven Magee
- Solar Photovoltaic Design for Residential, Commercial and Utility Systems by Steven Magee
- Solar Photovoltaic Operation and Maintenance for Residential, Commercial and Utility Systems by Steven Magee

26. Author Contact

Steven Magee,

3618 S. Desert Lantern Road,

Tucson,

AZ 85735

USA

I hope that you found the book informative and please let me know about any questions or comments about the book.

I am a consultant on new solar photovoltaic projects, solar photovoltaic system troubleshooting, solar photovoltaic training, and solar photovoltaic investing for financial companies. Please feel free to contact me for any help or assistance in these areas.

You may find my other books useful:

- Solar Photovoltaics for Consumers, Utilities and Investors
- Solar Photovoltaic Training for Residential, Commercial and Utility Systems
- Solar Photovoltaic Design for Residential, Commercial and Utility Systems
- Solar Photovoltaic Operation and Maintenance for Residential, Commercial and Utility Systems

www.ingramcontent.com/pod-product-compliance
Lightning Source LLC
Chambersburg PA
CBHW040826180526
45159CB00001B/84

* 9 7 8 1 4 5 2 8 1 9 3 3 4 *